# THE FLIGHT OF THE MONARCH AND OTHER REFLECTIONS

ALSO BY MICHEL BRAUDEAU

*Retour à Miranda*

*Six eccentriques*

*L'Interpretation des singes*

*La non-personne, une enquête*: "L'un et l'autre"

*Pérou*

*Loin des forêts*

*Esprit de mai*

*Mon ami Pierrot*

*Le livre de John*

*L'Objet perdu de l'amour*

*Naissance d'une passion*

*Fantôme d'une puce*

*Passage de la main d'or*

*Vaulascar*

*L'Amazone*

# THE FLIGHT OF *The Monarch* AND OTHER REFLECTIONS

BY MICHEL BRAUDEAU

*Translated from the French by Sophie Hawkes*

GEORGE BRAZILLER / NEW YORK

First published in the United States of America in 2004 by George Braziller, Inc.

Originally published in France in 2001 by Éditions Gallimard under the title *Le Monarque et autres sujets*

For information, please address the publisher:
George Braziller, Inc.
171 Madison Avenue
New York, NY 10016

Library of Congress Cataloging-in-Publication Data

Braudeau, Michel, 1946–
[Monarque et autres sujets. English]
The flight of the monarch and other reflections / by Michel Braudeau; translated from the French by Sophie Hawkes.
p. cm.
ISBN 0-8076-1530-7
1. Monarch butterfly—Migration. 2. Animal migration. I. Title.

QL561.D3B7213 2004
591.56'8—dc22 2003024564

Designed by Rita Lascaro
Printed and bound in the United States of America
First printing, May 2004

*for Eric Fottorino*

# *Contents*

*No Subject Too Small* 9

*The Glory of the Monarch* 15

*In the Toad's Embrace* 39

*In Praise of the Gnu* 53

*The Swift Sleeps on the Wing* 67

*The Ageless Turtle* 81

*The Flea* 95

*Postscript* 107

# No Subject Too Small

Nonjournalists often disparage the craft of the journalist, imagining him as a necessarily superficial person, who lies or misrepresents the truth and exercises an inordinate amount of power. (Until now, journalists have been too busy or too individualistic to take the initiative, but imagine what would happen if one fine day the profession decided that there would no longer be any newspaper, radio, television, or Internet news anywhere on earth, for a whole week. The silence would be deafening, unbearable even to our detractors . . . ) I consider this job, especially with regard to reporting, as one of the most magnificently free professions one can practice. As far as I'm concerned, it's impossible to imagine a job that is as thrilling, as varied, as demanding without being boring, or that arouses as much curiosity by opening as many doors in a legitimate fashion. A reporter fulfills his childhood dreams, like an artist, writer, or director. Some days it's hard, but he cannot really imagine what else he would want to do. Certainly not rest.

The word *reporter* came into French unchanged from the English and can be replaced, if one wants, by *enquêteur* (investigator), or *envoi special* (special correspondent). It all comes from the same source: writing down something one has seen; going someplace others cannot go, and giving news of it. And this is where the reporter finds his greatness as well as his modesty: his field of study is infinite. This is of course at times overwhelming, at times almost ungraspable. There is no subject too small for investigation. Mozart composed a staggering aria on a lost pin in *The Marriage of Figaro*. Similarly, one could write a fascinating investigation of a dog run over by a car, a demonstration by one labor union or another, a mailsorters' strike, or the horrors of war in Sierra Leone or Chechnya—by taking various risks, of course, and by following one's temperament, taking the angle most favorable to one's personal talent. But you must be convinced that each day a good part of the world is waiting to be discovered.

For a long time I wrote about other people's books, both French and foreign, the cinema, the Cannes and Venice film festivals. When *Le Monde* suggested I become a reporter, the idea appealed to me immediately, as an even broader field of investigation. And I've never regretted it, knowing that one year I'd travel around the world, and the following stay in Paris, and that from one time to the next it would never be the same. The offer to follow a few migrating animals during the summer of 2000 was another windfall. Migrating animals are very interesting creatures inasmuch as they are forced to travel, to change environments in order to eat or to reproduce. The monarch butterfly undertakes an extraordinary odyssey to follow its favorite poison and to sleep in Mexico. The toad migrates only a few dozen yards, but in just as vital a fashion: it's an amphibian and its eggs are laid in water before living on solid ground. The same reasons for travel apply to the common swift, the leatherback turtle, the gnu, and many others. I took the sturgeon out of this collection because my sources could not agree on its ocean journey. And I added a few pages on the flea, which was not part of the series originally published in *Le Monde*, because the flea is "erratic" and does not migrate like the others. But the flea, for reasons unclear to me—but no doubt related to William Blake's famous painting—is close to my heart. I wrote about it because it moves about in leaps and bounds, weightless and unconcerned about thoroughness. I trust it will forgive me.

I'd like to thank everyone who helped me in my research, beginning with Mr. Guy Jarry and his co-workers at the Muséum National d'Histoire Naturelle; Adolfo Castañón, an excellent writer and friend; as well as all those whom I've cited in these pages, especially Messrs. Eduardo Rendon-Salinas and Homero Aridjis, in Mexico.

Michel Braudeau

# The Glory of the Monarch

It's a king. A tiny king measuring a few fractions of an inch, at best, and weighing no more, even after a feast of nectar, than .028 ounces. A far cry from Louis XVIII, the overfed king. In spite of the solitary ring of its name, the monarch is not lonely; it lives and travels by the million. And yet its crown and its life hinge on very little: a weed, a few careless woodcutters. The monarch is a butterfly, not even the most beautiful of butterflies—the competition is stiff—a mere spark of orange and black against the blue sky. Yet, as we shall see, the fate of the monarch, whatever it might think—if it's foolish enough to think of its fate at all—is political.

We don't have the monarch in European countries; at most we have a few stragglers blown in by a hurricane or coming over by boat, but they don't live here. Its preferred territory extends from southern Canada and the northern United States to Peru, as well as the Marquesas, the Îles de la Sonde, and Australia. We don't know why it's called "monarch," although its distinguishing qualities make it worthy of the highest rank. It got its scientific name from the Swedish botanist Carolus Linnaeus (Carl von Linné, 1707–1778), who must have been having gastric problems that day, since he baptized it *Danaus plexippus.* Danaus was one of several sons of King Belus, scarcely illustrious in ancient myth, while Plexippus, an even more obscure character, was a hunter of wild boar, known only through a single verse of Swinburne. There are certainly more prestigious godfathers. As Jo Brewer notes in her book: "It is ironic that Linnaeus should have chosen the names of two such miserable heroes for this remarkable insect who, in the order of Lepidoptera, holds a fairly major position on the evolutionary scale, while the far more distinguished name—Parnassius Apollo—was conferred upon a butterfly that in a number of ways is far more primitive."[1]

In France, one can only get a good look at the monarch in movies—occasionally in documentaries on the art channel, Arte, or, more often, in the films shown at la Geode du Parc de la Villette and especially the Futuroscope Theater in Poitiers. At the Futuroscope, on two screens the size of several tennis courts, one in front of the spectators, the other beneath their seats, you can see a short but spectacular Japanese film on the monarch's migration from the Canadian border in October to the middle of Mexico, where they hibernate. You see it fly over Niagara Falls, lakes, and all the other dangers implicit in such a long journey: 2,600 miles in a month, one way, at an average altitude of 10,000 feet—a record among insects. But we'll come back to the monarch's migration later, the distance of which is hardly its most extraordinary feature.

If you want to see the monarch in optimum conditions, you have to go to Mexico City and get used to the altitude (7,800 feet). The air is thin and exceptionally polluted. From there you can choose an organized excursion or, much more interesting, get to know a local entomologist and let him be your guide. Someone like the young Dr. Eduardo Rendon-Salinas, from the Instituto de Ecologia UNAM, who often frequents one of the monarch's hibernation sites, high up in the mountains, two and a half hours northeast of Mexico City by car. You have to leave early to reach the village of Angangueo, world capital of the monarch, where everything is under the sign of the monarch, from the hotel and café signs to the souvenirs, to the shape of the jam cookies served for dessert at a remarkable cantina on the calle Nacional.

Angangueo is at the heart of thirteen monarch sites or sanctuaries, only five of which are really protected. From the village you take a pickup truck to climb the ten miles of extremely bumpy hairpin turns, not recommended for people with heart trouble. The rest of the trip, in the sanctuary proper, in this case El Rosario in the Sierra El Campanario, is done on foot, slowly (you are 10,000 feet above sea level), and preferably in silence.

At first it's like Hitchcock's *The Birds.* You see a monarch out of the corner of your eye, then two. Nothing to worry about. You are too busy counting your heartbeats and reading the small, none-too-exact, instructional signs nailed to the trees. Then you see ten, resting on the ground. Then a hundred, immobile on a branch. A thousand, around a footprint filled with water. As you proceed up the path, between the sacred firs (*Abies religiosa*), you see a huge orange cloud swirling around in the sun in one clearing after another. In the dark passages, the trees, tree trunks, and branches appear to be covered with a metallic, dark green armor. These are the black outer edges of the folded wings of the butterflies, which sleep one against the other, covering the entire surface of the trees. Here is one of the ways to estimate their approximate number: knowing that an adult monarch—and there are no young butterflies hibernating, only adults—weighs about .028 ounces, and that a branch of sacred fir is a certain length and the trunk a certain height, etc., and that so many pounds are necessary to bend a branch or to break it, one can deduce the population of monarchs present by dividing that weight by .028 ounces. In particularly opulent years, people estimate that by the end of October and the beginning of November, almost 5 million arrive. It is the only place in the world where one can distinctly hear the sound made by the wings of butterflies: when hundreds of thousands of them are flying around you, it sounds like the music of a light, fine rain.

Sometimes it's difficult to proceed without stepping on this continuous carpet of fragile wings, a concern that most woodcutters in the area don't share, but which one cannot help but feel if one comes as a visitor, especially when accompanied by a biologist. The solution is simple: you bend over and blow toward the ground. Detecting the presence of carbonic acid in the air, the monarch thinks it senses a predator, a feline, and immediately takes flight in large clouds. Clearly it's a problem when there are too many visiters in the sanctuary, or when children scream as they play and breathe too hard; then the monarchs move to the other side of the mountain.

Sometimes they move for other reasons. For example, inclement weather (which is pretty rare) and man have cut down many trees, allowing too much sun to come in, favoring the growth of an underbrush that is inhospitable to the butterfly and letting in the cold and the wind, which disturb the sleepers. But for the time being, and no one knows how long it will last, the thirteen or so Mexican sanctuaries represent a safe, cool, humid, and peaceful haven for the monarch, ideal for their winter sleep. They form a large, well-hidden quilt at a few mountain summits in a single state in Mexico, the state of Michoacán. The people in the area have known about the winter presence of the butterflies from time immemorial. Since the monarchs arrive in huge numbers more or less by All Saints' Day (November 1), they are believed to incarnate the souls of the dead, come to greet their families, and they are heralded in the Plaza Angangueo with fireworks and orange and black costumes. For Canadian and North American entomologists, however, the existence of this perched dormitory was an enigma, a terra incognita on the map. And to the scientific world the discovery of the site in 1975 by Fred A. Urquhart and Kenneth Brugger was a revelation on the level of that of the city of Machu Picchu by Hiram Bingham in 1911.

Until that date, the scientific community knew that the monarch migrated. But where to? The ones from the Pacific and Australia were easy to follow. The ones from the American Northwest ended up in California. But people would see those from Canada, Maine, and Vermont head south and disappear into thin air, God knows where. Fred A. Urquhart, professor at the University of Toronto, has nurtured a passion for the monarch since childhood. Along with his wife, he had already devoted many works and a long series of learned articles to it, when quite suddenly he got the idea of attaching a small label, indicating the place of origin and the date, to the butterflies' wings before they flew off. After persisting for thirty years, he attained his goal. The following year he published his findings in *National Geographic*: "Found at Last: The Monarch's Winter Home." Then he retired, hailed as the new Christopher Columbus of entomology. Of course this science, which usually develops into a passion, never lacks disciples. Today Dr. Lincoln P. Brower of Sweet Briar College in Virginia is considered the preeminent authority. He travels frequently to Mexico, where he has met with a succession of Mexican presidents and ministers of environmental affairs, who politely listen to him with more or less deaf ears. He meets with researchers like Eduardo Rendon-Salinas, who comes to the monarchs' sanctuaries many times a week in winter and enters his findings into a computer, to work on them the rest of the year. He represents the new generation of entomologists devoted to the monarch, those who know the most about this mysterious traveler.

The reason our understanding of the phenomenon of the monarch was so long in coming is due to its surprising complexity. The eggs are laid on a certain type of Asclepiad (known more commonly as milkweed), which grows abundantly on the Canadian border. This member of the Asclepiadaceae family is itself a toxic plant; its milky sap is poisonous, which protects it from the huge appetite of herbivores. But not the monarch, which over the course of time has become accustomed to the poison, to the point where now it can no longer do without it, at least in its early infancy. And the benefit it reaps is that the poison becomes lodged in the hard parts of the butterfly's wings, the exoskeleton, making it unsuitable for consumption. A bird that attempts to snack on a monarch immediately vomits and goes into cardiac arrest, which sends a clear message to its peers: don't touch the little orange and black guy. This is a useful protection when you're a butterfly lacking a beak and claws.

The egg remains between three and four days at 86°F, for eight to twelve days at 64°F, then becomes a handsome caterpillar, striped white, black and yellow, which devours Asclepiads for ten to thirty-eight days, depending on whether it's hot or cold outside, and attaches itself to the plant through acrobatic contortions, turning into a cocoon for nine to fifteen days, after which it emerges and spreads its brilliant wings. This individual born in the Canadian autumn will gambol about for a bit, gathering nectar from flowers. But bad weather comes on quite quickly. Even if the monarchs are intrepid creatures equipped with a sort of antifreeze that allows them to tolerate temperatures down to 17°F, they cannot endure the ensuing deep freeze of winter at -22°F. Therefore, they make provisions, fatten themselves up, putting on as much weight as possible, and head south. A wise move, for even if they had the means to resist the north, they wouldn't have any food, which itself dies off in winter. So off they go.

The journey lasts about a month, with an average of 45 miles a day, although records have been noted up to 215 miles a day. Wind, of course, plays a large role. The monarchs fly southwest to Texas, then head due south to the middle of Mexico, into the area of the sanctuaries, where they arrive around All Saints' Day, setting themselves up in colonies on the sacred firs. During migration they fortify themselves with the nectar of flowers found along the way. Once hibernating, they do not feed on the trees upon which they rest. They live on their reserves, in a state of lethargy. Around mid-February, the days become sunnier and the temperatures rise. You can see the monarchs flying in the light. This is not to find food, as the flowers still have no nectar—only pollen—but since the air has become less humid, the butterflies are in search of drops of water. Then, at the end of winter, the butterflies mate and leave the sanctuary. Most of the males die on the spot. The females, who have a more distant mission to fulfil, head for home, recrossing the Gulf of Mexico to lay their eggs on the Asclepiads available in this season in the southern United States (*Asclepias humistrata* or *viridis* or *asperula*), and die in turn. One or two generations of monarchs are born in the U.S. South in the spring, and the second heads farther north. Two other generations will be born in the Great Lakes area during the summer, feeding on another plant (*Asclepias syriaca*), which is plentiful at that time in northern Minnesota and along the Canadian border. That's where the fifth generation in the monarch's cycle will be born, the one that will come into being in autumn, migrate to Mexico, and become the first in a new cycle.

In other words, the parents, Mr. and Mrs. Danaus I, who migrate south in October and spend the winter there, will die separately, he in Mexico, she in either Texas or Louisiana. Their children, the Danaus II, and grandchildren, the

Danaus III, will be born, mate and die in the southern and central United States. With the help of warm weather, their great-grandchildren, the Danaus IV, and great-great grandchildren, the Danaus V, will return to Canada, where the Danaus I began. The family cycle thus comes full circle. The monarch has returned. Except that he or she is no longer the same, but an already distant descendant. A rather lax family, true, not terribly concerned with the incest taboo, but their numbers are so great, their lives so short, that one has to be tolerant.

We note, clearly, that the generation that flies south makes a much longer journey than those that head north and take four generations to return. The first, or "migrating," generation, lives for eight long months, while each of the following four generations live an approximate span of five weeks. The former is also unusual in another respect: most of the life of the "migrants" is spent in diapause, in which their sexuality is as if suspended and they cannot reproduce. It is not until the end of their

Mexican winter, when the temperature rises in February, that the butterflies begin to secrete the hormones that will lead the males to fling themselves irresistibly at the females, who are ready at last, and merge with them indissolubly in flight, four wings joined together in a position that leaves the human observer perplexed and envious. The subsequent, short-lived generations are immediately ready for reproduction and do not travel long distances. But we'll not extrapolate on the virtues of the asceticism and chastity necessary for the accomplishment of great things.

One of the most perplexing enigmas in the migration of the monarch butterfly lies in their return over several generations. Since a butterfly that is born in the north in autumn and will soon migrate is but the great-great grandchild of the last one to make the same journey, how does it know what direction to take, what precise moment to depart, and, most of all, how to find a sanctuary it has never seen, 2,500 miles from its native fields? People talk of genes, instinct, magnetic fields. It is also possible that the butterfly understands that it must migrate because it senses the winter coming, eagerly puts on weight like other migrating creatures, and heads south, attracted by warmth and the sun. And it is possible that the Caribbean and south winds lead it to the middle of Mexico as if through a funnel, where it comes across sacred fir forests that it finds absolutely charming, as its ancestors did before it. But when one thinks of the distances crossed, the dangers of the voyage and the narrowness of the sanctuaries, one senses that the answer is more complex. Lincoln P. Brower thinks that monarchs have something like a compass and a map inside them, which allow them to navigate to the other end of the earth, into the great unknown, and that their instruments are inscribed in their genetic code as soon as the first egg is laid.

According to Eduardo Rendon-Salinas, the monarchs appeared about 40 million years ago. But they didn't always migrate. First there were important climatic changes, and then there was man's appearance in North America. In cutting down the forests, man made way for the Asclepiads, which grow in destabilized areas. And if there were not so many of these plants, there would not be such a large monarch population. The life cycles of different sorts of Asclepiads and the monarch are intimately connected; the butterfly can only be born on this plant, and in turn favors its dissemination. Depending on the point of view, either the insect follows its food source, or the plant promotes its best pollinator. In any case, the vastness of the monarch population ensures its survival. A heavy snowfall can kill millions of subjects at a single blow. Many drown in pools of water, sink into mud puddles, or fall from their trees, stone dead, for reasons still unknown. While the milk of Asclepiads gives monarchs their brilliant colors and protects them from most predators, it does not discourage all hungry creatures. Such is the case with the calandra lark, which understands the danger found in the hard parts of the butterfly and is careful not to swallow them, boring a hole into the butterfly's abdomen with its beak, then using its tongue perversely to empty the abdomen and absorb its healthy vitamins. The finch contents itself with eating it in small quantities, so as to avoid succumbing to the poison. It takes a few bites, stops when it feels that it is about to vomit, moves on to something healthier for a moment or two, then returns to its butterfly for another risky nibble. When hibernating, monarchs are a bit chilled and slow and don't always escape the terrible black-eared mouse, which is apparently insensitive to the butterflies' poison and blissfully devours their abdomens, where they store their provisions of fat, by the thousand.

But these little alimentary massacres are nothing compared with the damage caused by the monarch's greatest predator, man, who doesn't even have the excuse of wanting to eat it. The sanctuaries are threatened by the deforestation, either authorized or clandestine, which has struck over the past few years. The discovery of these sites by scientists in 1975, and the consciousness-raising that followed, more or less coincided with an increase in the human demographics in the region, resulting in a surplus of population in need of new lands to cultivate. The last fifteen years have seen the rise of a sort of guerrilla war between the companies that exploit the forest and local poverty on one side, and ecologists, scientists, and artists on the other. The latter contingent have come together in the group known as the Hundred, with the poet Homero Aridjis as their spokesman: "I was born in Contepec, in the state of Michoacán, near the sanctuary of Altamirano. When I was a child, every year we went to the mountains to see the monarchs. We did not know they came from Canada, any more than the Canadians knew they came from here. Every time I went to see the tombs of my relatives I saw monarchs. They formed a large orange river in the sky around All Souls' Day [November 2], creating a sort of natural mythology. One day in February 1985, the pollution level in Mexico City became unbearable. You felt like you were dying. The next day, I read a letter from a philosopher friend of mine protesting the lack of efficient measures against this endemic problem. I called him, suggesting we demonstrate, and we contacted other friends, Octavio Paz, Gabriel García Márquez, Juan Rulfo, the painter [Rufino] Tamayo, etc. This is how the group of the Hundred was born, and how I found myself the head of an ecological movement. The forest where the monarchs go is a microcosm, a protective wall against the sun, a humid refuge. If you hack your way into it, you're cutting holes in a roof, with the same dire consequences."

With great determination, the Hundred pressured the Mexican government, which issued a decree on October 9, 1986 prohibiting deforestation in the thirteen sanctuaries during the hibernation period. "It's absurd, like outlawing the breaking in of houses while their inhabitants are there and authorizing it while they're away. . . ." Only five sanctuaries are completely protected, officially at least, since clandestine cutting does exist. "We have nothing against the peasants, to be sure. They are very poor and remain so. They get forty dollars for a tree that took sixty years to grow. In its place they plant corn on sloping land, which is hard to cultivate. The harvest is bad and brings in very little. To cut trees to make firewood or fruit-crates is a stupid waste. They could live off the forest by keeping it intact. They simply need to learn sylviculture. But nobody wants to teach the peasants anything. . . ."

Aridjis is not a gentle poet sensitive to little flowers and pretty butterflies. His greatest struggle is against pollution in Mexico, especially in Mexico City, where the levels are staggering and tied to the political system. But, like others, he thinks that the beauty of the monarch contributes to the beauty of the world, to its necessary diversity, even if it is not directly "practical." In California, almost simultaneously, people have become alarmed over the possible disappearance of the monarch's hibernation sites, ceaselessly eroded by the growth of suburbs and highways. The Los Angeles County Natural History Museum's Monarch Project, developed by Chris Nagano, helped lead to the passage of a law protecting the species in 1987. The American essayist Diane Ackerman (*A Natural History of the Senses*) was won over by the project. In her book published in 1995 (*The Rarest of the Rare: Vanishing Animals, Timeless Worlds*), she devotes a chapter to the "The Winter Palace of the Monarch" and describes the diversity of these "winter palaces" in California: a beach house in Santa Barbara, a youth camp in Big Sur among bears and coyotes, a motel in Pacific Grove, a eucalyptus forest not far from Hearst Castle, etc. It seems that when migrating to California (which concerns only the monarchs born west of the Rockies), the presence of the ocean is an element favorable to hibernation. The Californians have a sense of the celebratory, and in some sites they welcome the arrival of the butterflies with an orchestra, clowns dressed in orange and black, local poets reading their works, and the hoisting of a banner with the colors of the monarch, which is only taken down after their departure. In Pacific Grove, anyone caught disturbing a monarch, in any way whatsoever, is subject to a five-hundred-dollar fine. But here, like everywhere else, the sites are threatened by human expansion: highways, golf courses, riding ranches, camping sites, and luxury villas, competing for the seaside. The Monarch Project tries its best to fight the progressive destruction of these sanctuaries.

And why? Because, as Diane Ackerman writes, this insect has something uplifting about it: "They are silent, beautiful, fragile; they are harmless and clean; they are determined; they are graceful; they stalk nothing; they are ingenious chemists; they are a symbol of innocence; they are the first butterflies we learn to call by name. Like the imagination, they dart from one sunlit spot to another."[2] Entomologists add that there might be other reasons to want to protect the monarchs, reasons not totally aesthetic and to some extent abstract. It has recently been noted that monarchs that accidentally taste transgenic corn die from it. It might be interesting to note that perhaps we, too, should be more careful about what we eat. As for Lincoln P. Brower, he thinks that if we ever truly succeed in understanding the monarch's navigational system, we might be able to use it to guide space probes into still uncharted skies.

# In the Toad's Embrace

There is a winding county road in the Chevreuse Valley that joins Cernay-la-Ville to Auffargis, passing by the romantic ruins of a Cistercian abbey in whose midst arise the splendors of an out-of-the-way hotel and a curious modern art installation. For three hundred yards, on either side of the road, sheets of doubled white plastic, half buried, hang from posts two feet off the ground. In the morning sun, it looks like a corridor of white light, one of the landscape interventions that American artists dubbed "Land Art," whose most spectacular representative is perhaps Christo, who wrapped the Pont-Neuf. The Christo of the Chevreuse Valley, a thirty-year-old named Grégoire Loïs, is conducting zoological studies; he is working not for the sake of art, but for the love of toads. How this athletic man, with arms like girders, came to be conquered by the common toad, the tiny triton, and the agile frog we shall see below; the fact remains that he sincerely loves them.

Just what hasn't been said about the poor toad? Jean Rostand recalls its legendary dual nature in his work, now unfortunately out of print, *La vie des crapauds* (The life of toads): "It suckles on cows, makes wine turn, pilfers from birds' nests, lays waste to beehives, has the evil eye, casting spells on man and beasts alike; it dies if stared down; gives rabies to dogs with its slime; its breath is poisonous; it sullies everything it touches." Yet at the same time there's always a chance the toad might turn into a handsome prince: ". . . It cures gallstones, dries up dropsy, stops nosebleeds, soothes pain . . . keeps rats away; and sometimes one finds a golden stone with magical powers under its head." Even the great naturalist the comte de Lacépède couldn't contain himself when discussing the toad in his

*Histoire générale et particulière des quadrupédes ovipares et des serpents* (General and specific history of oviparous quadrupeds and snakes): "dirty in its dwelling, disgusting in its habits, deformed in its body, murky in color and pestiferous in breath; it opens its hideous mouth when attacked, with the stubbornness of a stupid creature". At the end of this eloquent arraignment for the crime of homeliness, he quite simply proposes the extermination of the species. Luckily there are people who are moved by the beauty of the ugly.

Such as Grégoire Loïs, who noticed that during the month of March, at a certain spot in the Chevreuse Valley on the banks of the pond in les Vallées, there were hundreds if not thousands of toads, both male and female, crossing the road at the same time. They would form a viscous carpet on which cars would skid and sometimes end up as part of the scenery, squashing a large number of the batrachian population in the process. He had the simple and efficient idea of putting up plastic sheets to block the toads' way, planting buckets in the ground into which the little creatures would fall. Then he needed only to collect them and carry them to the other side of the road, saving their lives and allowing them to reproduce in the pond. Once their mission in the water is accomplished, the toads come up against the same obstacle on the return trip; Loïs or other volunteers then carry them back to the forest side of the road, counting them and noting their sexes along the way. And they do this every day from the end of February to the beginning of March, during the period of migration. For the toad migrates.

Of course it's nothing like the monarch's imperial migration, but rather an insignificant sort of migration, a mile or two at most, from the underbrush to a stagnant pond, from one lowly spot to another. But it is indispensable to the survival of the species, for the toad is an amphibian that lives in the open air as an adult but is born in water.

Named *Bufo bufo* by Linnaeus, the European common toad is a tetrapod (it has four legs) and anuran (or tailless, unlike other batrachians such as the salamander). Its skin is very permeable, which affords it a great deal of exchange with its environment. Varying in color from brick red to earthy green to olive, the skin on its back is warty, covered with vesicles containing a poison that is excreted only in violent situations, such as when a dog bites. The white poison, which is thick and toxic, contains an alkaloid that makes the dog salivate for hours and cures it of any penchant for toads. They say that certain aging hippies in Australia, no doubt bored with LSD, lick toads' backs when they want to hallucinate.

The male is small and smooth, the size of a frog. He's skinny and moves in short jumps when he moves at all, since, truth be told, he doesn't move that much. The female is a lot bigger, and she's the one you see in gardens and gazebos, eating grass. How long do toads live? Jean Rostand, who kept a lot of them, had one for thirty-six years, and even then it succumbed only to a Labrador retriever belonging to one of his assistants. But such a long life is not that unusual. The toad has a very slow metabolism, like reptiles or bats, which can live up to thirty-three years, unlike its enemy the vole, which is always agitated, races around at over 100 miles per hour, and dies young. If it is protected from predators and does not have to forage for food, as was the case with Rostand's, the toad has no stress and lives for a long time. In reality—and we know this from sawing a toad's bone and counting the circles indicating the annual changes in rhythm—a toad left to its own devices lives about seven years.

The European common toad lives between sea level and an altitude of 6,500 feet; from oceanic climates to the very tip of the Balkans; in cultivated areas or in woodlands; and near man. In the past, people used to sell them at flower markets because they get rid of slugs and eat earthworms and ground beetles by projecting their sticky tongues, attached to the fronts of their mouths, like chameleons. The toad always lives in damp places. Laid as an egg in the beginning of March, it lives in a larval state in the water. As a young tadpole it has internal gills for two months, through which it filters oxygen from the water. Then one day it wakes up without a tail, with four legs and lungs. It leaves the water, by now a valiant little toadlet. But it is not yet time for the joyful love feasts. It must mature first.

The male reaches maturity in its fourth year, the female one year later. Daily life among toads varies according to climate. In summer it keeps fairly busy. With the coming of winter, those in the south continue to enjoy life, like the Provençaux. The ones in the north must move to their wintering sites, discreetly, slowing down their biological rhythms. The choices are simple: either you move more and more in order to find food or simply tighten your belt and wait. Or, of course, you die like insects, leaving your eggs behind. The toad chooses to tighten its belt and sleep.

At winter's end, the drama abruptly begins. Whether you're from the north or the south, it's all hustle and bustle, all hands on deck, and without a bite to eat, everyone races to the reproduction site, meaning the local pond. The whole toad population within a six-mile radius dashes to the same spot, at the same time. Hence the carnage along the roadside, and the protective plastic. The toads have no thought for cars at a moment like this; what really bothers them is the sexual ratio of the species: there are many more males than females. The sexual competition heats up quickly; all females will be engaged, but not all the males. Many males will remain on the sidelines, like the Cervidae (the deer family), contenting themselves with watching their chief, the best of the lot, sporting with the most beautiful females. Since the reproduction strategy is to lay as many eggs as possible, this requires a lot of water so everyone can jump into the bath at the same time and try his or her solitary luck in this aquatic version of Sodom and Gomorrah.

How does the anxious male resolve the problem that he and others of his gender outnumber the females? By his wits. The shrewdest choose their females before setting off, settling themselves on their backs and holding on very tightly with their forelegs, riding their big females to the water. Pirate toads try to take their places or, through nearsightedness, bad faith or blunder, mistake them for females and climb on top. The first occupants push them off with their strong hindlegs and emit brief "croaks" of protest. Rostand writes prettily: "They often sing during coupling." No doubt. But beforehand, while courting. The female tree frog, for example, is more attracted to the males that sing the best. And this is not for love of music; it's a secondary sexual signal that indicates an important genetic trait: the singer over there will have the best immune system, for example. But during the great journey toward coupling, this "croak" is not singing; it means, "I was here first, hands off my girl."

Other males, who think themselves more subtle, wait until the females reach the pond to pounce on them. Finally, once in the water, the orgy begins, and a female can find herself assailed by many males at once, and it's no joking matter. There is no penetration, as the male has no penis. Both sexes have only one hole, used for everything. The male grips the female with his forelegs in the posture known as "amplexus" and digs his fists in under her armpits, both to hold her and to help her, over the course of several hours, to expel the two gelatinous ribbons, the oviducts, which contain the eggs, which he in turn sprays with his sperm. The surrounding males can spray them, too, as in porno movies. There is no jealousy and perhaps no pleasure either. Rather, it all corresponds to a very strong need.

The males, who do absolutely nothing all year long, know that today they have a great mission to fulfill. Their whole progeny is at stake. Some are seen gripping pieces of wood, or bottles. Sometimes they kill the female by suffocating or drowning her. It's next to impossible to make the males relax their grip. Rostand cites experiments conducted by a certain priest named Spallanzani. Father Spallanzani took a male weighing .9 ounce coupled with a female of 3.7 ounces and had to attach a weight of 7 ounces to her before the male would let go of her. Twelve times his weight. Acrobats will appreciate this. He cut off both legs of another toad (we'll spare you the details of the torture carried out by this sadistic prelate), without being able to force it to let go, while in the meantime the maimed male fertilized the eggs. So, Mr. Lacepède, don't you think such tenacity is worthy of a noble Roman?

After the battle, the predators—polecats, shrew mice and grey herons—come to gather up the dead. But the greatest predator is the return journey. The females, having laid some six thousand eggs or so, are no longer of interest to any of the males. The eggs are left in the water, abandoned to their fate, except among the type of toad known as the "midwife toad," which watches over them and protects them until they hatch. The others go home to rest, taking their tortuous mysteries along with them, especially their lopsided sexual ratio. Why are there more males than females, whereas, as far as reproduction is concerned, one male is enough for several females everywhere else in nature? And furthermore, wonders Grégoire Loïs, why the sex? "In nature, sex is dangerous . . .".

Not everybody likes the plastic installations. Sunday cyclists, on their titanium bicycles, dressed in their loud, fluorescent cycling garb, call it aesthetic aggression and would turn a blind eye to the extinction of all toads. Loïs shrugs it off. He has his own ideas about the thinking biped, pedaling or otherwise. When it came time for his military service, he was a conscientious objector and was thus required to serve double time, on a stipend of two thousand francs a month. So he had to find an odd job. He did so in the kitchens of the hotel in the nearby abbey, and that's how he first saw the toads migrate. And get squashed. Since there are many conscientious objectors in environmental protection organizations, he later resumed his studies in biology, earned a DEA (Diplôme d'études approfondi—roughly equivalent to an M.Phil.) degree, and joined the Muséum Nationale d'Histoire Naturelle. He then contacted the regional national park, which became interested in his ideas. The ideal solution would have been to dig tunnels under the road for the toads, but that costs money. They preferred using big hearted volunteers, free of charge, perhaps hoping that one day the toads would change ponds. Digging a substitution pond on the other side of the road wouldn't help a thing, since toads are stubborn. "They are very attached to their traditional reproduction site," says Loïs, "and they return there even if it has dried up in the meantime. In that case they die there. Then one day, an adventurous toad, a 'deviant,' a Christopher Columbus among toads, finds a new site and starts a new migration, which, with a little luck, won't cut across a county road. It's always the deviants that assure the continuation of a species."

# In Praise of the Gnu

If there ever was a wild animal that gets short shrift from humans, surely it is the gnu. What's a gnu, if not a butt of silly wordplay? Oh yes, the gnu, like in the song "*I'm a G-nu, how do you do?*"[3] Or people remember a particularly bloody documentary they saw on television. The gnu? That idiotic creature that lets itself be devoured by the hundredfold in the river where crocodiles feast on them? What a dunderhead. Some animals are a little stupider than others, aren't they? Let's call a gnu a gnu. Even the name is ridiculous; it's not pronounced like "gnome," which resonates a bit, or like "gnostic," where the sound is modified into "nostic"; it's not oiled, as in "gnocchi." "Gnu" sticks to the palate, like a little lump, then does a U-turn, giving gooseflesh to the mouth. Anglo-Saxons solved the problem with the alternative, Dutch-derived *wildebeest*, which is a slight deformation of *wild beast*—the English are on the ball, once again. Yet here we should read "wild" as "feral," as opposed to "ferocious." The gnu is no doubt untameable, but nothing if not timid.

The *Petit Robert* French dictionary informs us that the word *gnu* is of Hottentot origin, which is heavy with connotations if you think of the massive and patently unseductive silhouette of the Hottentot Venus, a true damper of desire. And the *Petit Robert* kindly adds that it is "an ungulate mammal from South Africa, with a heavy body, thick, shaggy head and slender legs, an antelope's trunk, the head and horns of a bull, the tail and mane of a horse." In other words, it looks like nothing at all; it has no specific personality, and is nothing more than a freak of nature, a poorly stitched together patchwork of debris cast off by various congeners and unable to coexist in any other combination.

If one is to begin to do justice to the gnu, one must go to Tanzania and Kenya, where the gnu abounds in a group of nature preserves almost one-tenth the size of France, about 34,000 square miles, notably in the Serengeti National Park and the neighboring Ngorongoro Conservation Area, between Lake Victoria to the west and Lakes Manyara and Natron to the east. Usually one sets out from Arusha, a tiny town east of the Ngorongoro Crater, which serves as a base for safaris. The Ngorongoro Crater, for its part, is situated on the upper bank of the Rift Valley, the gigantic geological depression that divides Africa and stretches from the coast of Mozambique in southeast Africa to the Dead Sea and the Jordan River valley. Long ago, the Ngorongoro was a powerful volcano that must have reached 16,000 to 20,000 feet, ever spewing out lava and ashes. The prevailing winds from the south and southeast, the monsoon winds, rained down ashes to the west of the volcano for millennia, in successive layers, over millions of acres. The volcanic covering filled in the cracks in the relief like a great blanket of asphalt, allowing only a few rocky peaks, the isolated tips of former mountains, to jut out, providing refuge to a few trees. From these heights, one can admire the immense Serengeti Plain, flat as a billiard table, the "endless land" of the Masai.

But the Ngorongoro became tired, coughed less and less, collapsing into itself two and a half million years ago, and is now a mere 8,000 feet high. The crater became a cauldron, a caldera, with a flat bottom over 12.5 miles in diameter, home to elephants, gazelles, buffalo and zebra, practically everything except humans. The latter can, however, nest in one of the hotels perched over the void, which are very impressive, and it isn't unusual to spot big, fat tourists in these vertiginous sites, in broad daylight, who would make a meal for several lions. Sheathed in brightly colored, synthetic sportswear and shod in futuristic athletic shoes, these tourists take turns having themselves photographed in front of this Dante-esque landscape, at a safe distance from the wild animals. From up here, the animals are hidden by trees, and the lush vegetation is shrouded in fog. One sees only the mosquitoes and the baboons, naked and dignified.

Only when descending the Ngorongoro toward the large plain does one begin to see Thomson's gazelles, giraffes, and zebras. The acacias are the only trees to grow in the layer of soil covering the shield of lava, their umbrella-like foliage slight and almost without shade. Nothing obstructs the view, except for a few undulations of the terrain, hills. No one lives here, outside the little tourist lodges, except for the park staff, the conservators, and the guards who drive all-terrain vehicles and know the animals' habits. You can't drive down there at just any time, anywhere, to pick a flower or sprinkle a few drops on the parched earth, since there is a strong chance the scrub might conceal one or two sandy-coated lionesses, invisible for the time being, who with a swipe of their infallible paws can mow down your flower and everything else along with it. The guide will tell what places can be visited at what time, and let you see the cats up close without danger. You simply cannot get out of the car, though you can leave if the window is open. Since they are inured to the frequent visits by the Land Rover, or because they don't distinguish you as separate from the whole "noncomestible vehicle," they don't pay any attention to you. And they have no shame, either; if you come during a week when the lioness is in a good mood, the lion won't have any scruples about honoring her right before your delighted eyes, two yards from your tires.

Further on, at the edges of a salt lake, vultures circle over floating, bloated, or beached carcasses with ribs pointing skyward. These are gnus that have drowned, explains Japhet, our unsentimental guide. So here are your first gnus, carrion torn apart by large, bald-headed birds. Farther on, past the several hundred fleeing beige gazelles with white rumps, almond eyes, and twirling tails—reminiscent of eternal virgins, in spite of the abundance of the species—

you see the zebras. Aside from the fact of always having freshly painted stripes, the zebra has the good luck to look very clean in any circumstance. It can roll in the dust and get up spotless, mane like a brush and buttocks bouncing.

Whenever you see zebras, according to Japhet, you are also bound to see gnus, sooner or later. Here, certain animals fall in with one another willingly. The ostrich, for example, with its powerful legs and its skull perched atop its long neck like the head of a nail, has very keen eyesight, but poor hearing. The antelope has mediocre vision, but very good senses of smell and hearing. The two animals travel together. Such is also the case of the zebra and the gnu, which are both fine runners and vegetarians that get along very well: the zebras, with their strong teeth, go first and eat the tall, hard grass. The gnus, less well-equipped, come next and reap the short, more tender grasses. So they form a team, each with its own nature, the zebra with its energetic independence, the gnu with its stubborn submissiveness.

Suddenly you see one or two gnus. Not too clearly individuated, dark and dusty with a furtive and disheveled look and a bulky silhouette, they turn on their heels and disappear. Japhet is confident: it's the beginning of May, we should see plenty more. Plenty. He gets his information by wireless from mysterious correspondents. They have spotted columns of gnus to the south, a few hours from here; the bulk of the herd should come tomorrow. For the great migration of the gnus has begun, like every other year since the beginning of time.

Gnus migrate in search of food, as do many other species. Gnus eat all sorts of grasses, but always young shoots (less than 4 inches high), not only because they are tender, but above all because they are richer in the minerals and proteins that favor the birth of well-formed baby gnus. The young shoots appear three weeks after the beginning of the rainy season, which starts south of the equator and continues first to the north, then to the south. Thus do the gnus follow the rains and their daily fare, from the south to the north and back again, in a circle about 930 miles in circumference. They can spot lightning in the sky hundreds of miles away and sometimes move toward storms that never break. According to Professor Pierre Pfeffer, the migration of the gnus in the Serengeti is the last such migration of large animals that one can still witness, unlike the bison in America or the springbok in South Africa, which ended up succumbing to the bullets of white colonists.

The next day the gnus are indeed plentiful. From the top of a hillock, you see the black armies come over the horizon, slowly, kicking up dust and lowing mournfully. You cross a hill and suddenly, as in those films of historic battles when hordes of Austrians or English appear out of nowhere, you discover hundreds of them on a slope. A little farther on, thousands. These numbers are not exaggerated; as many as 2 million gnus have been recorded migrating at once. Pierre Pfeffer cites the case "of an English zoologist whose Land Rover was blocked by a congestion of gnus covering almost 40 square miles."

In three days, he had all the time in the world to observe his jailers. Gnus are large animals, 4.25 feet high and weighing between 440 to 550 pounds, but they're not very good-looking. Mr. Pfeffer admits, "If, judging from its behavior, the gnu's intellectual faculties are far from brilliant, its physical appearance is scarcely more winning; as a zoologist I have always wondered why my distinguished predecessors classed it among the elegant antelopes." Its coat is dark brown, vaguely striped at the shoulders; its bearing lacks grace; the gnu is heavy in the front and low in the back, which makes it look cowardly. As for the modest intellectual capacities to which Mr. Pfeffer refers, they are translated into a spirit of submission not far from utter stupidity. The gnu follows a leader, usually an older female, in whole herds, even if the leader is on the wrong track. If a zebra suddenly takes charge and goes in a different direction, say east instead of north or west, toward Lake Victoria, as we once witnessed, thousands of gnus will fall into step without missing a beat. If the leader wants to cross the unavoidable Mara River on the Kenya side, as has been the case for millennia, at the same inconvenient spot where crocodiles unblinkingly await their daily fare in the water—to say nothing of the lions, panthers, and hyenas on the opposite bank—all the gnus rush into the river at the same time. They don't necessarily get across the river, but they all jump in, one after the other, and they sink, suffocate, and drown by the dozen. The crocodiles have a field day. And the survivors pay toll once again when they emerge from the water. On the average, a lion eats thirty-six gnus per year. This week it's a free-for-all, without surprises, for the gnu wouldn't think of disobeying or breaking rank. It's nature's cannon fodder.

The gnu's only answer to these preordained mass slaughters lies in its great fertility. The females do not bear their young at different times of year, but all together. The time of the dropping of the young is triggered by hormonal factors related to the quality and supply of the grasses. The mothers can stall birthing for up to four weeks, while waiting for favorable foodstuffs. At the time of reproduction, for a period of three weeks, eight thousand gnus are born each day. The females drop their young communally, without taking the trouble to imprint their young with their scent (which, regrettably, results in many of them getting lost). The young are on their feet within a few minutes and run as fast as their mothers in less than an hour. If not, predators would eat them immediately. And in fact, predators do eat quite a few of them right on the spot. But no matter how hungry the lions are, at some point they do have to take time out to digest their meals. Eating gnu day in and day out can get tiring. So, in spite of the massacres, the gnu population remains stable at about 2 million heads, perhaps not thinking heads in our opinion, but heads perfectly adapted to their environment nonetheless.

In the end this large though unexceptional beast, this humble foot soldier, does have something endearing about it in spite of itself. Its modest way of existing without persecuting its neighbors, accepting the laws of its species without complaint—it's touching. In a way the gnu has no say in the matter. It is a nobody. No nation or kingdom has put the gnu on its arms, its blazons. One finds lions and eagles aplenty. Even unicorns, though imaginary, not to mention ostriches, roosters, and llamas. But no gnus. It's a good sign. And its alleged ugliness is relative. Individually the gnu is graceless, nothing like a pink flamingo. But if you see it run in a herd, jostling sideways and dancing out of step like a drunk, with its big, dark head that looks as if drawn by a bad-tempered child, like a clown's head with curved horns, ears like oars and a sage's little gray beard—a very modern look, all told—the gnu can inspire one of those irrational feelings of affection that last a very long time.

# *The Swift Sleeps on the Wing*

At the mouth of the Gironde, across from Royan, the lands of southwestern France become drier and flatter, narrowing into an acute triangle, called Pointe de Grave, a slender tongue of sand and pines and a few small dunes where the French Navy maintains a base and has an elegant, ancient semaphore. This is where the ferry lands after crossing the Gironde, where you first see campers fleeing the overpopulated right bank in summer, and where, in the spring, you can see the most beautiful migrating birds. The LPO (la Ligue de Protection des Oiseaux, or the Bird Protection League), which has its headquarters in Rochefort, in the Corderie Royale, sends ornithologists to the site on a regular basis. They set themselves up on the dunes, equipped with binoculars and notebooks, and document the passage of birds. Joined by volunteers, they are long on patience, facing southward for more than two months, from mid-March to the end of May, from sunrise to sunset. According to Olivier Maigre, a young ornithologist in the field, the first swift of the year was spotted at la pointe de Grave on April 7, 2000. By May 2 they numbered 3,026; by the 6th, 33,611 and by May 17, the total was around 52,000 subjects.

La pointe de Grave is a place of passage frequented by a great many migrating birds—more than two hundred fifty thousand are counted each year, including one hundred thirty different species. They do not like to fly over water, where they cannot land in case of need and cannot find anything to eat. Thus they take advantage of the tongue of land as much as possible, so as not to cross the Gironde River until the last minute, regrouping themselves in a bottleneck as in certain mountain passes in the Pyrenees, the most famous of which is Organbidexka, the "free pass," so called because there are no hunters. In spring, one sees, pell-mell, turtledoves and honey buzzards, barn swallows and house martins, chaffinches, goldfinches, and, more rarely, cuckoos, kites, and storks. One also sees common swifts, about which less is known than the others. The turtledove is familiar, storks bring babies to the Alsatians, the cuckoo is the patron saint of marriage in France, but the swift? It is neither here nor there. You can see it, but cannot catch it. You can hear it, but it is already gone. Indeed, the common swift, a dazzling but discreet artist, is the only bird to share with the albatross the cumbersome privilege immortalized by Baudelaire: "Its giant's wings keep it from walking."

The swift is often confused with the swallow, which it resembles. Like the swallow, it has curved and pointed, scythe-shaped wings. In addition to the pallid swift and the Alpine swift, there is also the common swift (*Apus apus*), the one that we're interested in. It looks black from a distance, seen from below, and dark brown up close. It measures 6 to 6.5 inches in length, with a 16- to 19-inch wingspan. In France you can see them chasing each other in flocks in the evenings, cheeping in summer, above the rooftops in towns, always up high. There are a lot of them in Montmartre, for example. What distinguishes the swift from all other birds is that it devotes by far the largest part of its existence to flight. According to Guy Jarry, an eminent specialist on the common swift, among other birds, and biologist at the CRBPO (Centre de Recherche sur la Biologie des Populations des Oiseaux, or Center for Research on the Biology of Bird Populations), it is estimated that in fifteen to eighteen years of life, by flying three hundred miles a day for its simple activities of personal comfort, the common swift travels more than half a million miles a year.

Suffice it to say that its migratory flight is not a big undertaking for it. This journey stretches from the Arctic and the north of Europe in summer to Africa in winter, as far south as the equatorial forests such as in the

Congo. Like many birds, it follows the insect population. The swift is among those that arrive in France the latest and leave the earliest. It captures its food in flight, which includes coleoptera, diptera, and spiders dislodged from their webs by the wind. The swift has considerable, unchanging dietary needs. In the period of reproduction, which is to say during its European summer, it is capable of abandoning its nest and its young for hours, searching for food over hundreds of miles, if necessary. Its incubating eggs and young are capable of resisting extreme heat, placing themselves in a kind of suspended animation, lowering their body temperatures to conserve energy.

The swift lays, once a year, up to three eggs. The females sit on the nests, where the males join them. It's one of the rare moments in a swift's life when it rests. But not for long, since incubation lasts about two weeks. The young are abundantly fed, stuffed to plumpness, then summarily abandoned. The maturation of the chicks coincides with the exhaustion of their accumulated fat. Then they manage to leave the nest on their own.

Thus the great number of false starts. And a false start is fatal for the common swift. It has enormous wings and ridiculously short legs—it is even classed among the "apodidae," the animals lacking feet. So it cannot right itself to raise itself up and take off. It's all over once it hits the ground. Unless it's thrown into the air, like a kite. Or it's lucky enough to fall into the hands of certain volunteers such as Mme Effinger, in Poissy, not far from Paris, who rescues and tends to dozens of endangered swifts each year.

Originally it was a cliff bird, later taking advantage of human constructions to leave its craggy environment to come and live in towns. That's why it nests up high, on rooftops. In order to take flight, it needs an incline to plunge into the void. Once airborne it stays there for several days. To feed itself it simply opens its mouth. The mouth is pretty when shut, pointed and streamlined for speed, but horribly large when open, a funnel for insects, which makes it a little like the well-named *engoulevent* (the Eurasian nightjar, whose French name literally means "wind swallower"). To drink, the swift skims the surface of the water, using the lower half of its beak like a spoon, the same principle used by Canadair firefighting planes. The swift gains from being seen from a distance. Up close it's scary: the English nicknamed it "devil's bird." "It's always the same when man comes across a black animal," notes Mr. Jarry. "If it were white, it would be in a museum showcase. But black . . . And with that gigantic mouth. Furthermore, we don't know what it does at night. It's mysterious." At night, among other things, it sleeps.

But the swift sleeps on the wing, of course. It flies way up high (and can even reach twenty-six thousand feet when it chances to spot, owing to its extraordinarily acute vision, a cloud of insects trapped in a rising hot-air bubble) and glides as it sleeps, first with one eye then the other, a few moments at most. The swift glides a great deal. Its albatross-like fuselage allows it to take advantage of all ascending currents, riding the wind like a fine racing boat. It soars and it roves. "Its challenge is to be a city-dwelling aerial insect-eater. Unlike swallows, which avoid large polluted, densely populated cities such as Paris, the Parisian swift goes looking for food in the forests of Sénart or Fontainebleau. When you can fly 50 to 60 miles an hour, it's no big deal." Furthermore, gliding is energy-efficient, unlike wing flapping, which uses a lot of energy and forces the bird to stop to refuel frequently, thus exposing itself to danger.

Great fliers are for the most part chaste. While seventh heaven may be found between the sheets, the deed is pretty quickly done in the real heavens. "One hole touches another," specifies Mr. Jarry, "with an injection of spermatazoa. They do it in flight. Do they repeat it? I don't know. It takes only a fraction of a second; it's extraordinary." That's certainly one way to approach it. Is the swift faithful? Nobody knows. To be absolutely certain, you'd have to band the birds with radio transmitters, but the right kind of equipment would be too heavy for them.

The swift has practically no predators. No raptors are fast enough to catch it. It has too many resources. It's vulnerable only when nesting, and this is why it is very careful when changing sites, studying the lay of the land for three or four years before making a decision. The swift, generally speaking, is very faithful to its nesting site, which can work against it; tiny flies have been noted in the nests before the birds take off, and they remain in the larval state during the winter, reviving in the spring to burrow into the swifts' auditory canals, fatally unbalancing them. Friends of the bird thus regularly destroy its nests, for reasons of hygiene, and the birds willingly rebuild them in the same place.

The only predators known to it are the children of the Central African Republic, who, when the termites hatch in the fall, attach the fat insects to long, fine threads which the swifts swallow like fishhooks. There aren't enough termites in Point de Grave for children to play this cruel game, but we have something better: hunters. The hunter of yore was a single male, more or less camouflaged, who caught birds in nets and then fattened them up. Or he would wait patiently up in his hide for the passage of migrating birds, making his own cartridges, which took a certain amount of time and limited damage. He would come down at lunchtime and reward himself with a fine rib of beef. When the time came, he would vote for the prohunting candidates. Today the hunter is less peaceful and more stressed out. His cartridges are mass produced and cost a few cents each, which makes him generous at the moment of attack. An unpredictable zoological phenomenon, the French gendarmes, who had previously limited themselves to citing hunters only in places where they weren't often found, such as on public thoroughfares, have begun to venture into the underbrush. Is the policeman undergoing a genetic mutation? Or is it a political mutation?

Anyway, the hunter, an endangered species, has tried to cozy up with politicians most antagonistic to human migrations, but nothing doing, the gendarmes keep handing out fines. But not too many. They have to catch the hunter at the moment he is pointing his gun at a protected bird, in other words, when one can say, beyond the shadow of a doubt . . . Nonetheless, this grave insult to the Tartarins of Medoc[4] has led them to demonstrate in front of police stations and mayors' offices. No mayor in the Medoc is elected without the hunters, and people will eat doves in southwestern France and ortolans in Latché for a long time to come.

Of course, the swift is protected by law, like the turtledove. But accidents do happen. Round la pointe de Grave, one constantly hears rifle shots, more than a thousand in the early morning, and one doubts they are very selective. Allain Bougrain-Dubourg comes every year to protest and invariably exposes himself to verbal abuse. After all, what could be more normal than the changing of the seasons and migrations? In the past, the protectors of birds, the ornithologists, received death threats. Today they're merely insulted. Apparently the question of age-old tradition is being raised. The tradition of killing for no reason. "After all, in many countries, torture is a tradition. Should it be continued for that reason?" asks one of the volunteers present.

In this month of May, the human species here divides into two categories, those from above, from the dunes, the airy people, and those from below, the loudmouth earth people. A bus stops at the foot of the dune, and a rowdy group gets out. As soon as one of the earth people spots a bird, he shouts, "Bang! Bang!" as if hunting. His friends find this very funny and follow suit. They are from the Medoc. Rugby players. As we approach, we see that they are in fact a class of boys about to take the ferryboat to play a match on the other side of the river. So young, and yet already convinced that "it takes two guys from Charente to make one from Gironde," among other macho aphorisms. You might ask why the birds persist in flying over the ilk of them. The ornithologist answers, "Because they don't learn to mistrust. It would have to enter into their genetic makeup and that takes millions of years. And the dead ones tell no tales."

# The Ageless Turtle

On the coast of French Guiana, driving northwest, you leave horrible Cayenne, with its leprous colonial houses, corrugated metal, and infernal heat, and drive northwest. The man at the car rental warned you not to leave anything visible inside the car, since in their exasperation crack addicts will steal anything. The excellent road, which is quite luxurious for this inhospitable overseas department of France, passes by Kourou, the Îles du Salut (Islands of Salvation), which include the Îles du Diable (Devil's Island), where Dreyfus was imprisoned while waiting for Zola to make lightning strike in France. You don't see any rockets in the air, but you know the flawless, handsome Ariane spacecraft are there.

Later the road becomes pocked with potholes, the farther you get from Sinnamary. You go through Jojo, Trou Poissons, Trou Caïmans, Dégrad Savane and indeed the road continues to deteriorate all the way to Mana, where there's a faded, almost illegible sign indicating, at the very bottom, the direction to Awala-Yalimpo and the beach of les Hattes, terminus for the curious and a few other large beasts that run aground there, equally exhausted.

The beaches of les Hattes, Awala, and Yalimpo are the traditional destination of the leatherback turtles, *Dermochelys coriaceae*, known as *tortues luths*, or "lute" turtles. The phenomenon of their yearly return is so predictable that the turtles are even mentioned, in parentheses, on the map. And in fact, you can see signs of them even by day, since the beach is turned completely upside-down by these powerful and silent musicians, who, as soon as the sun sets, never miss their appointment, staying all night long, until daybreak. The French call them "lute" turtles because their convex backs are divided into seven long folds, which makes them look like lutes or inverted mandolins. But they do not sing like whales or dolphins, and are in fact totally mute. And their shells cannot be strung for guitars once they die because they don't have shells.

Aside from other peculiarities, this turtle is a soft turtle. Not soft like a soft Dali clock, but lacking the plates of armor that protect its cousins from raptors, if not from the makers of eyeglass frames. Nor does it possess the bony box into which other cousins withdraw their heads and feet at the slightest danger. The *Dermochelys coriaceae* is exposed and vulnerable. It has a skeleton on the inside, over its internal organs; its back is thickly coated with two inches of fat that is quite solid, but flexible; its skin is a beautiful dark blue, spotted white, and looks like leather (hence their English name, leatherback). Its thick layer of fat allows the leatherback to go into much colder waters than other marine turtles, and its flexible back allows it to grow larger than other turtles confined to a hard shell. On the average it measures between six to seven feet long, and weighs between 500 and 800 pounds. One turtle once weighed in at more than 2000 pounds.

During the laying season, the female does not eat, and as her layer of fat diminishes, her ribs begin to show. We're familiar with the turtle when it's on the beach, and we know about its laying habits. What it does in the water, on the other hand, is anybody's guess. They have been tagged since the 1960s, with metal or plastic bands which they quickly lose. In Mexico people mark hard-shelled turtles by removing a scale from the white belly and grafting it onto the dark back of the same turtle, thereby removing the risk of rejection; depending on the placement of the light scale, one can tell the provenance or the date of the operation, etc. That's not possible with the leatherback, since it has no scales. Recently, scientists have perfected small electronic markers called PITs, passive integrated transponders, which are like bar codes in supermarkets. They are injected under the turtle's flexible back, then read by means of a handheld receiver with an antennna, indicating the place and date of the "pitting." Otherwise, we know that the leatherback reaches sexual maturity between thirteen and twenty years of age, and that a female leatherback lays around a million eggs during a three-month period, from April to July.

But what does she do outside the laying months? The CNRS (Centre National des Recherches Scientifiques, or National Center for Scholarly Research) has perfected radio transmitters known as Argos platforms (*balises Argos*), and harnesses that don't impede the turtles' movements and come undone after a certain amount of time, freeing them. After leaving French Guiana, some leatherbacks swam to Cape Verde and Senegal. Others went to Florida or toward Labrador. We know their reproduction and feeding sites. Everything else is vague. The leatherback, a charming eccentric, eats only jellyfish, which are 90 percent water. So it needs to eat a lot of them, even though a reptile doesn't have the same energy needs as a fish and has no central heating. The turtles gather together in the Perthuis Charentais, an inlet on the western coast of France, well-stocked with jellyfish. But it is also stocked with plastic bags, which the leatherback, unsuspecting and, nearsighted as a footstool, mistakes for its favorite food, causing a fatal obstruction of the bowels.

The leatherback, a marine reptile, breathes. As it approaches the water for the first time, it does not adopt the undulating movement common to reptiles, but rows the sand with its great-fingered flippers. It breathes through its lungs, but scientists suspect that it also benefits from an oxygen-storing system in its mucous membranes. For while it can easily hold its breath and dive 1,000 feet, returning to the surface to breathe every five minutes; it's also capable, if necessary, of plunging to 3,300 feet in a quarter of an hour, no small feat.

Between Awala and Yalimpo stands the building of the Mana Nature Reserve, among the coconut trees near the beach. Thirty-year-old Vincent Liardet, former worker for the Kawana Campaign launched by the World Wildlife Fund for the protection of turtles in 1992 and 1994, coordinates a small group of young volunteers associated with an Amerindian institution, Kulalasi. The wildlife preserve is spartan in its comforts; one sleeps in hammocks hung inside a screened enclosure where the mosquitoes bide their time, digesting. Outside, after 6:00 p.m., it's a perpetual "Mosquito-blitz" with famished squadrons on constant patrol, reducing men to a permanent state of gesticulation, grimaces and masochistic slaps, recalling a madhouse or a Louis de Funès skit. Over two miles of beach and thirty miles of coastline, the team engages in scientific observation and marks the females. This is how we know that a turtle returns seven times in a season to lay hundreds of eggs. A large part of the team's work consists of sensitizing the public—tourists and children—to respect the turtles. The male turtles are very discreet and keep their distance. One never sees leatherbacks coupling, even though we know that the male is equipped with some sort of penis he must use when out of the reach of the waves. The female, who does all the work, as usual, is equipped with a sperm bank, a reservoir that allows her to create several series of eggs without soliciting her inconstant partner. To tell the truth, the whole scenario is a bit speculative, since it's impossible to observe it firsthand. The leatherbacks spend a tiny fraction of their existence on the beach. With the Argos platforms, people have begun to follow them on the high seas, but very little to date.

We don't know how long leatherbacks live, for example. It is tempting to think that they live to be worthy centenarians like other turtles, but it hasn't been proven. We've never seen an intermediary specimen between the baby turtle and the adult, which leads one to think that they grow rapidly. What then? On the other hand, we do have proof of the leatherbacks' fidelity to their reproduction sites in the Atlantic, in French Guiana and along the coasts of Africa today, and formerly, in Indonesia,

Malaysia and Mexico in the Pacific, where they were exterminated in the nets of Chilean fishermen. In any case the leatherback is a born swimmer, crossing oceans effortlessly, reaching speeds of thirty mph, navigating who knows how, by sense of smell or by the stars. It is believed that they have crystals inside their heads that make them sensitive to the nuances of the earth's magnetic field and that they also use the smell of deep-sea currents.

According to Vincent Liardet, it is not easy to construct the turtle's ethnology. "For every thousand eggs laid, only one will become an adult turtle. This gives an indication of the role the turtle plays in its environment as a food source. The eggs are attacked in the sand by mole crickets and large burrowing insects during the two weeks of their incubation. After they hatch, crabs and stray dogs devour the tiny newborn turtles. Those that make it to the water fling themselves into the arrays of catfish waiting for them, half grounded on the seashore, mouths open wide." At the same time, other interested parties come from above: night herons, vultures, great horned owls. On the open sea, the white shark and the tiger shark follow in hot pursuit, depriving the leatherback of its useful limbs. On earth, sometimes a jaguar emerges from the underbrush and bites off the turtle's head, a scene worthy of Delacroix, though man, however, has so far succeeded in filming only the unfortunate results of such an encounter.

Erosion is also a factor. Some turtles, too tired or stupid to know better, lay their eggs too low on the beach, in areas assailed by the sea. To say nothing of certain nights when several hundred turtles arrive at once and lay their eggs together, some in the nests of others, thereby suffocating each other's eggs. Let's finally admit it: the leatherback, a majestic beauty, is not exactly brilliant. It apparently has less initiative than an old turntable. No matter what, it carefully digs its laying hole thirty-five inches deep, alternating shoveling movements with its left and right back flippers. Even if these flippers were amputated it would make the same movements. And if it runs up against an obstacle on the return trip to the sea, such as a fallen branch, rather than going around it, the turtle simply butts against it until dying of sunstroke.

Daniel William, unofficial Amerindian chief of the Awala region, devoted to the protection of the species, says that the turtles represent nothing so much as ocean spirits to his people. He also says that in eating leatherbacks one risks becoming as stupid as they, something not to be taken lightly. The Amerindians have always eaten leatherback eggs, which are rich in protein, but in just proportions. Today, poachers steal eggs on the beach and sell them for forty cents each in Suriname, on the other side of the Maroni River. To get at them more quickly, they cut open the belly with a knife, like a lid, and carry off a hundred or so eggs, enough to buy themselves a few doses of crack and blow out their fine, superior, biped brains. Other leatherbacks, the ones still out at sea, get their heads caught in the meshes of Surinamese fishing nets, which are several miles long, and drown there.

There are only twenty-five thousand turtles remaining in the area. Still, the leatherback could one day become the subject of a very coveted and precious study for mankind—from an egotistical perspective, of course. People have noticed that the leatherbacks do not die of old age, but rather of wear and tear. Which is to say they die because they are eaten, drowned, eroded, or wounded, but not old. In man, the cells divide and progressively lose information. With the leatherback there is no such loss. At the time of its death, its chromosomes contain the same enzyme as in the beginning; the DNA segment remains perfectly intact. You might see images of wounded and dead turtles, but you won't see any of old turtles. If Hollywood could only solve this mystery...

It's not the fate of the leatherbacks that most worries Daniel William, when he worries. He is more afraid of the growth of North American religious sects that are destroying the indigenous social fabric of the Amerindians by giving them "free" assistance, cash gifts, and the famous good news of the saving of their souls. "We would be happy without the Whites," he explains, "the French included. We don't care about rockets or satellites. We have only one television station, and mobile phones don't work here."

# The Flea

The flea does not migrate, but it travels. It's an erratic creature. Scholars have long debated its origins and fabulists its qualities. It has even been argued that it crossed the ocean with the Vikings. The flea doesn't leave much in the way of fossils, as you can imagine, but it does leave something. Today, according to Professor Jean-Claude Beaucornu, one of the most distinguished specialists on the flea, who teaches in the medical school at the University of Rennes, we are pretty sure that the human flea (*Pulex irritans)* originated in an area that includes the southern United States, the isthmus of Panama, and northern Latin America. All fleas came from this area, and most have remained there, except one, *Pulex irritans*, which crossed the Bering Strait when it became a land bridge after the waters receded in ancient times. The flea traveled in an animal's coat, such as, perhaps, that of the common fox, one of its favorite hosts. This all took place around the end of the tertiary period (perhaps several million years ago), well before the Vikings. The oldest known *Pulex* fossil dates from 4,500 years BCE.

* * *

How long does a flea live? It belongs to the class of Holometabola, insects like the butterfly. First you have a little larva, then an inert pupa in which the insect takes form, then the mature insect, the adult flea. If the conditions for the development of the larva cease to be favorable, it goes into diapause, a state of suspension. Let's imagine that the nymphs are brought into an apartment by people who then leave for several months. The flea nymphs go into diapause, their development arrested five or six months, as long as it is cold or food is lacking. They wait for the humans, and the heat, to return and then, with the first noise—the slightest impact or stimulation suffices—the diapause is broken and the fleas emerge from their nymphs, in just a few seconds. Similarly, the nymphs laid in the cushioning of hummingbird nests will wait for six months, the whole span of the annual migration, to be born when the hummingbirds return, when the birds reuse the material from their old nests to make new ones. Otherwise, entomologists estimate that in nature a flea lives from three weeks to a month.

* * *

The flea lives alone, even if it shares its bedding with other fleas on a common host, where it reproduces. The flea is sexed. The genital apparatus of the male includes long erect antennae, "full of chemical and tactile receptors, which," according to Professor Beaucornu, "are easy to spot on the testicles . . .". The female flea burrows beneath the host's skin with an orifice for breathing and another for defecating and reproducing. The male flea copulates with her, thanks to a penis attached at a right angle.

The flea copulates only once in its lifetime. The female stores everything she will need for future fecundations in a sperm bank. "The larvae eat detritus and the *excreta* of the adults, 95% of which is made up of the host's hemoglobin," says Beaucornu. This transmission through stool engorged with the future host's blood accustoms the young flea to this food source. The fleas adapt themselves to their host; if they change hosts their reproduction rate drops.

* * *

Thousands of years ago, fleas lived on rodents or small, warm-blooded dinosaurs. They shunned the big ones. Monkeys did not have fleas. Man, as long as he was nomadic and lived by hunting and gathering, had no beds, and no fleas either. He acquired fleas when he settled down, in the shift from hunting-gathering to cultivation. He used to live in rock shelters, protected by piles of rock and fire. The cracks in the rocks allowed the flea-bearing rodents to come and go. After the glaciers receded, man continued to reproduce his former habitat, piling up stones into houses and fortifications. Rats and mice, along with their fleas, came from Asia to join him. Then he got genets, wild animals found on certain blazons of yore, then cats, which replaced the genets, and dogs.

* * *

Professor Beaucornu laments the decline of the human flea: "When I was in Algeria in 1961–62, you had only to shake a burnous for fleas to drop out of it. . . ." Today, you don't often find *Pulex irritans* around here. "Unless you go up to the drunks and ask them to give you their fleas, which they're not likely to do. In winter, when it's very cold outside, you can collect them at the hospital when the drunks end up there, if you're quick about it. Human fleas die in the hospital, where it's too warm, too clean." It's easy to find cat fleas, *Ctenocephalides felis*, which can withstand anything, but not the *Pulex*. In fact, since the 1950s, the *Pulex* has not been able to tolerate modern comforts. It likes cracks, rags, old boxes. Drunkards, who live under bridges outside, offer ideal conditions in terms of temperature and humidity: "*Pulex* loves them. . . ."

*Pulex* became accustomed to human blood, and is the only flea to have done so. It has an American cousin, which did not travel on the backs of foxes long ago, called *Pulex simulans*, which shares the same genome. When the Bering land bridge was submerged millions of years ago, the two cousins separated and evolved on opposite shores. Says Professor Beaucornu, "This included the shape of the male genital apparatus, which is the 'key in the keyhole type,' and no longer works. The cousins no longer have any affinity for each other. If they copulate in spite of this, since they're cousins and think they can, nothing happens."

* * *

In 1894, a French doctor, a disciple of Pasteur, discovered the bacteria that caused the plague in Hong Kong. Three years later another Frenchman, the colonial doctor Paul Louis Simond, brought to light the role of the rat, then that of the rat flea, *Xenopsylla cheopsis.* The plague was finally brought under control or domesticated. The plague of Justinian, about 540 CE, lasted two hundred years and caused 200 million deaths. The Black Plague, which left Asia in 1340, caused more than 25 million deaths in Europe, which is to say half the population of the continent between 1348 and 1350. In 1894, a third epidemic left the Indies, claiming 11 million victims by 1912. The numbers are staggering. Other less far-reaching epidemics have continued to crop up here and there. Civilization brought the plague to Madagascar, where it was previously unknown. Many scientists have been interested in the idea of spreading pestilential fleas over their enemies in wartime, by aerial means. This would simply repeat the ancient Tartar response to the Genoese siege, when the former used catapults to hurl the corpses of plague victims onto the latter, forcing them to withdraw.

In the sixteenth century, one approached plague victims only after donning a tight-collared, long leather chasuble coated with linseed oil, a hat, and a mask with a long beak filled with cotton imbued with sweet-smelling essences that were supposed to fend off "miasmas." In and of themselves, miasmas do not exist. The word *miasma* was used to designate harmful vapors emanating from the earth or from bodies, or provoked by the wrath of God in order to punish mankind. Thus outfitted, doctors used long pincers to touch sick people and to lance buboes; priests used pincers to give communion to the dying. People came to suspect that the plague was a contagious disease, without knowing about its sprightly vector. However, there are some cases, according to Professor Beaucornu, where the plague is spread without fleas, such as in Iran, where it is transmitted by small rodents, gerbils, that children catch and eat. "These kids all have scratches on their hands and inoculate themselves with the plague before bringing it home. An epidemic is unleashed, favored by the Muslim tradition of vigils for the dead: the fleas leave the man's body as it gets cold, and jump onto the people gathered to mourn him . . . ."

* * *

J.-M. Doby, a former professor emeritus of parasitology in the medical school at the University of Rennes, has collected some of the countless references to fleas in European folklore, beginning with the earliest songs, stories, and novels. The flea has always been closely associated with a woman's body, because it approaches her more quickly, more furtively and more avidly than her lover. The flea drinks in the virgin's blood while the suitor is still at the kissing stage. The French expression, *avoir la puce à l'oreille* (to have a flea in one's ear), indicates a very specific kind of itch. Girls like you to whisper in their ears. Everything the imagination associates with the female sex, such as secrecy, hiding-places, menace, and magic—danger or treasure—suits the small stature of the flea, as well as its bouncing, evasive manner of locomotion. If a man-sized flea could jump as high as the Eiffel Tower, one can well imagine just how powerless certain cuckolded husbands, simple pedestrians, must feel in the face of their elusive rivals.

The flea is like desire. It takes hold, and not always in the right place. You can cage it, as did Rabelais's Panurge, who wore a flea set into an earring. You can offer it jewels and gold chains. Or you can get rid of it by trapping it in ivory tubes or in furs, as did Duke Charles the Bold with the mink he wore around his neck, which had golden paws, ruby eyes and a diamond-studded muzzle. Mary Stuart, queen of Scotland, wore ermine. The women of Marseilles used to wear stoles of simple cloth, called *pistolets.*

* * *

Until the mid-twentieth century, flea circuses were popular in Europe, and spectators viewed them through magnifying glasses. One could admire the fleas pulling tiny carriages, jumping through hoops, or firing miniature cannons. History still preserves the names of the directors: Obicini, flea tamer under Louis-Philippe; the baron of Walkenaer during the French Revolution; or the famous Bertolotto, who performed first in London, on Regent Street, and then in New York, in 1875 in Union Square, with a troupe of one hundred artist fleas. Around the same time, a flea circus was in operation in Paris on the rue Vivienne. In 1923, a flea tamer named Mme Stenegry officiated at Place de la Nation in Paris, and in 1930 a Professor Tomlin set up his circus at the Olympia, in London. The fleas were neither tamed nor trained, strictly speaking. People simply artfully manipulated their reactions and attempts to flee. On the other hand, they were fed from the very skin of their "tamers," on the forearm in general, under bandages, to prevent them from escaping after dining.

You can see a flea tamer with his troupe on his arms in Orson Welles's 1955 film *Mr. Arkadin.* Tex Avery created a touching cartoon about the sentimental life of a flea circus, in which the heroine, Miss Fifi, talks with an unmistakeable French accent. But flea circuses always used human fleas, *Pulex irritans*, especially the female, for the acrobatic displays. The feline flea, which is less powerful, is not suitable for such shows of strength. Since *Pulex* is slowly disappearing from our countries, flea circuses have one by one been forced to close their doors.

* * *

It is said that among certain Polynesian peoples, where fleas were unknown before the arrival of the whites, fleas are considered supernatural beings and are respected as the souls of dead white men.

# Postscript

This small collection of essays on animals is my fourteenth and slenderest book. Prior to it, I wrote dreams, novels, fictions, and many portraits and newspaper articles. None has ever had the privilege of being translated into English and published in the United States. It is often said that Americans are not very willing to read translations from a foreign tongue. True, they have a rich and flourishing literature of their own, which I love and admire. I have, moreover, had the opportunity to do a great many interviews with authors as diverse as William Styron, Philip Roth, Susan Sontag, Joyce Carol Oates, not to mention, last but not least, that friend of France and my own, Jim Harrison.

I therefore express thanks above all to George Braziller, the patriarch, and all those who work at the publishing house he founded, since it is one of the few to manifest a sincere interest in non-American authors. Generosity in the realm of art and the mind is a gift that, fortunately, is never forgotten in the heaven of writers and publishers, strange insects that live on words and dwell in paper.

I would also like to salute those animals that have allowed me to approach them so I could get better acquainted with their often mysterious inner workings. Some have very short lives, while others, such as turtles and parrots, live a long time. Here I have focused on a few nondomestic, wild animals, with whom it is difficult to establish as long-lasting a friendship as it is with a dog or cat. All are migrators with serious, vital reasons for moving about, such as eating or reproducing, some covering a distance of only twenty-odd yards, such as the toad, others spanning thousands of miles, such as the leatherback turtle and the monarch butterfly.

And for travel, they resort to strategems that are sometimes quite complicated. Thus on the wings of a weightless butterfly was I able to cross the Atlantic for the first time.

Books wanting to convey a "message" are often pretentious and pointlessly moralistic. If this slim volume has any claim to ambition, it would be to get readers to show more respect and pay more attention to the animal world. Many authors, such as Diane Ackerman, and environmentally oriented nongovernmental organizations are already striving toward this end, more in the United States than in Europe, but it is still not enough. Developing for over two centuries, the industrial world has unfortunately brought about the extinction of numerous animal species, the famous dodo bird and the Tasmanian tiger, to name only two among hundreds. A great victory would be won over human arrogance and rapacity if we could make people understand that diversity of life on earth is an incalculable treasure (even from an egotistically human point of view); that certain animals need not disappear merely because they look odd to us or won't let us pet them; and that the humble orange and black butterfly in the Texas sky is an admirable, discreet athlete.

By ignoring them or allowing their destruction, we irremediably erode God's work (whoever this God may be to each of us), that treasure of Creation that was given us, generation after generation, at birth, for the brief time of our passage on Earth.

# *Notes*

1. Jo Brewer, *Wings in the Meadow* (Boston: Houghton Mifflin Co, 1967), pp. xvi–xvii.

2. Diane Ackerman, *The Rarest of the Rare: Vanishing Animals, Timeless Worlds* (New York: Random House, 1995), p. 139.

3. Flanders and Swann, "The Gnu Song," on *At the Drop of a Hat*, 1956–59.

4. Tartarin is a loud, egotistical character invented by Lucien Alphonse Daudet in *Aventures prodigeuses de Tartarin de Tarascon* (1872).